THE ELEMENTS

Chlorine

Susan Watt

BENCHMARK BOOKS

MARSHALL CAVENDISH
NEW YORK

Benchmark Books
Marshall Cavendish Corporation
99 White Plains Road
Tarrytown, New York 10591

Library of Congress Cataloging-in-Publication Data
Watt, Susan.
Chlorine / Susan Watt.
p.   cm. — (The elements)
Includes index.
ISBN 0-7614-1272-7 (lib. bdg.)
1. Chlorine—Juvenile literature. [1. Chlorine.] I. Title.
II. Elements (Benchmark Books)
QD181.C5 W35 2001
546'.732—dc21   2001025252 CIP AC

Printed in Hong Kong

Picture credits
Front cover James L. Amos/Corbis.
Back cover: Tracy Frankel/Image Bank.
Ann Ronan Picture Library: 18, 19 (*right*).
British Plastics Federation: European Council of Vinyl Manufacturers 22.
Brown Partworks: NASA 12.
Corbis: James L. Amos 30; Jeffrey L. Rotman 14.
Hulton Getty: 13.
Image Bank: Chris Close 23; David W. Hamilton 17; Peter Hendrie 27; Sandra Filippuchi 21; Siqui Sanchez 16; Tracy
Frankel *i*, 4.
Janine Wiedel: 19 (*left*).
Mary Evans Picture Library: 24.
Robert Hunt Library: 25.
Science Photo Library: Charles D. Winters 9, 10; CNRI 20; Juergen Berger, Max-Planck Institute iii, 15; James Holmes, Hays
Chemicals 8; NASA 6; Pat & Tom Leeson 26.
TRIP: A. Lambert 11.

Series created by Brown Partworks Ltd.
Designed by Sarah Williams

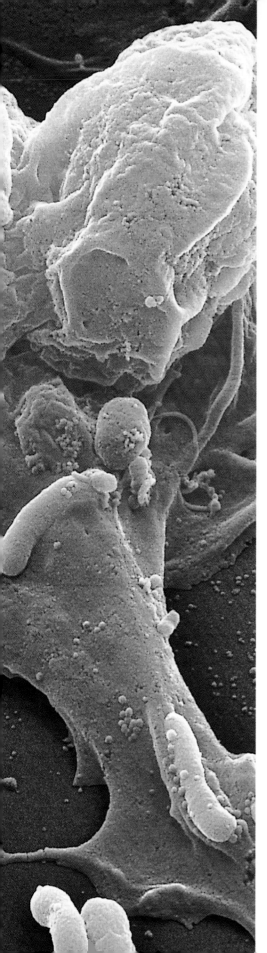

Contents

What is chlorine?

Chlorine is very effective at killing germs. Swimming pools are safe for you to swim in because chlorine is added to the water to attack disease-causing bacteria.

Chlorine is an element with two sides to its character. On the one hand, it is an incredibly useful substance that is contained in almost everything we buy and use—from the food we eat and the clean water we drink, to the clothes we wear and the computer used to write this book. On the other hand, on its own, it is a poisonous, choking gas, and many of its compounds can cause serious harm to humans and the environment.

The reason chlorine can be both good and bad is because it is so reactive. In fact, it is one of the most reactive of all the elements. In the periodic table, chlorine belongs to the group called the halogens, which also includes fluorine, bromine, and iodine. All the halogen elements react in a

---

DID YOU KNOW?

CHLORINE ISOTOPES

Every element has an atomic mass, which is the number of positive particles (protons) plus the number of neutral particles (neutrons) in the atom's nucleus. Chlorine's atomic mass is 35.5. This is not a whole number, because chlorine is made up of a mixture of two slightly different chlorine atoms, called isotopes. These both contain 17 protons and 17 electrons and are exactly the same chemically. However, 76 percent of chlorine atoms have 18 neutrons in their nucleus, and so have an atomic mass of 35, while 24 percent have 20 neutrons and an atomic mass of 37. Therefore, the figure of 35.5 is an average, taking into account the normal proportions of the two main chlorine isotopes.

CHLORINE ATOM

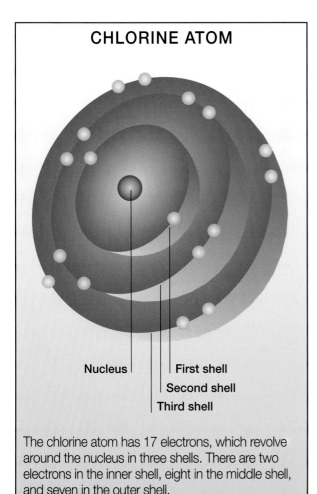

Nucleus | First shell
Second shell
Third shell

The chlorine atom has 17 electrons, which revolve around the nucleus in three shells. There are two electrons in the inner shell, eight in the middle shell, and seven in the outer shell.

ATOMS AT WORK

Chlorine has the symbol Cl. Each Cl atom contains seven electrons in its outer shell.

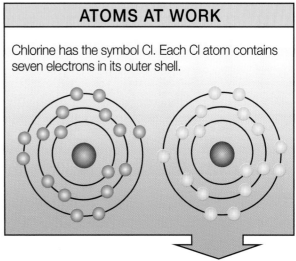

When chlorine is not combined with any other element, two chlorine atoms link together to form a molecule. They share an electron so that each atom has eight electrons in its outer shell.

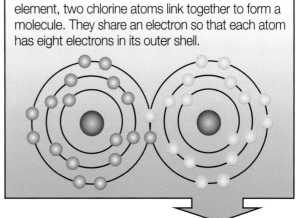

Chlorine gas is made up of these pairs of atoms, called diatomic molecules. The molecule's formula is written like this:

Cl_2

similar way and form a large number of compounds. Fluorine is the most reactive element of all, and chlorine is the third most reactive, after oxygen.

Atoms and molecules

Chlorine atoms all have 17 electrons. These tiny, negatively charged particles orbit the atom's heavy center (nucleus) in three shells—the inner shell contains two electrons, the middle shell contains eight electrons, and the outer shell contains seven electrons. This last shell is in effect incomplete, because it actually has room for eight electrons. This is why chlorine is so ready to react with other elements—in a chemical reaction it can gain or share an extra electron and fill its outer shell.

Where chlorine is found

Chlorine is found very widely in the world, but it never occurs on its own because it is so reactive. It combines with a whole range of different elements, but, most often, it is found combined with sodium as sodium chloride. This compound is better known as common salt—the substance we sprinkle on our food.

Chlorine is around the 18th most abundant element on Earth. Most of it exists as salt dissolved in the oceans and in salt lakes such as the Great Salt Lake in Utah. But huge amounts are also found in the ground as halite, or rock salt. Chlorine is also found in the human body, where it is the 10th most abundant element. About one part in 1,000 (0.1 percent) of our physical makeup is chlorine.

SEE FOR YOURSELF

GROWING SALT CRYSTALS

Sodium chloride occurs in nature as crystals of salt. We sprinkle tiny salt crystals on our food, but you can grow much bigger ones at home. With the help of an adult, pour some hot water into a clean cup or heatproof glass. Add a teaspoon of salt and stir the mixture until the salt dissolves. Add more teaspoons of salt in the same way (you can heat the solution up a little for a few seconds if it cools down too much). Continue until some salt remains on the bottom even after stirring. Then get a ruler and tie a length of thread around the middle of it. Put the ruler across the cup so that the thread hangs down in the liquid but does not quite reach the bottom. Then leave the whole thing on a shelf for two days. DO NOT stir the liquid or move the cup during this time. After two days, you should find a salt crystal formed on the end of the thread, and you can lift it out to take a closer look.

The Great Salt Lake in Utah, seen from space. The salt concentration is much higher in the water on the right side of the lake, making it appear bright blue.

How chlorine was discovered

Chlorine was discovered in two stages. Although Swedish chemist Carl Wilhelm Scheele (1742–1786) is usually credited with the discovery of chlorine in 1774, he was working at a time before people knew about chemical elements, so he did not realize that he had discovered a new one. He did realize, however, that he had produced an important new substance, and he set about recording its physical and chemical characteristics.

Scheele's discovery

Scheele produced chlorine by heating manganese dioxide with hydrochloric acid, which was then known as muriatic acid. According to theory at the time, combining a substance with oxygen (which was also unknown as an element) was called dephlogisticating. So, based on this theory, Scheele named the new substance dephlogisticated muriatic acid.

A generation later, British chemist Sir Humphry Davy (1778–1829) worked out that dephlogisticated muriatic acid was really a chemical element. In 1810, Davy announced his discovery to the world, and he named the new element chlorine, because of its color (*chlor* is a Greek word meaning "yellow-green").

ATOMS AT WORK

The reaction that led to Scheele's discovery of chlorine started with manganese dioxide and hydrochloric acid being heated together. This caused the bonds holding the molecules together to break.

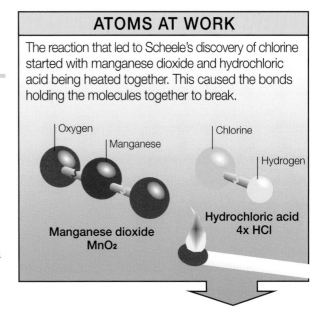

Oxygen

Manganese

Chlorine

Hydrogen

Manganese dioxide
MnO_2

Hydrochloric acid
$4x$ HCl

Some of the atoms then combined to form manganese chloride. At the same time, the oxygen atoms linked up with the hydrogen atoms, forming water. The two remaining atoms of chlorine joined together to make a molecule of chlorine gas, which escaped and was detected by Scheele.

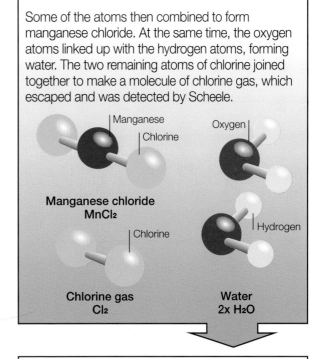

Manganese

Chlorine

Oxygen

Manganese chloride
$MnCl_2$

Chlorine

Hydrogen

Chlorine gas
Cl_2

Water
$2x$ H_2O

The reaction can be written down like this:

$$MnO_2 + 4HCl \rightarrow MnCl_2 + Cl_2 + 2H_2O$$

One molecule of manganese dioxide combines with four molecules of hydrochloric acid to give one molecule of manganese chloride, one molecule of chlorine gas, and two water molecules.

Extracting chlorine

Chlorine is extracted from seawater and from rock salt, in a process called electrolysis. In this process, an electric current is used to change chloride ions (produced when salt dissolves in water) into chlorine atoms. Electrolysis also provides other useful chemical products—hydrogen gas and sodium hydroxide, known as caustic soda—but it uses a huge amount of electrical energy.

In industry, three different methods are used to produce chlorine by electrolysis.

Each year more than 30 million tons (27 million tonnes) of chlorine are produced around the world by electrolysis in industrial plants like this one.

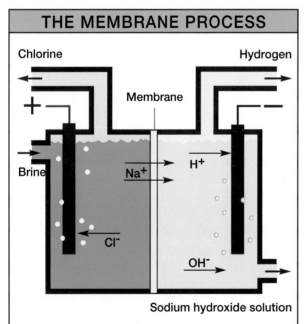

THE MEMBRANE PROCESS

Chlorine   Hydrogen

Membrane

+   –

H^+

Brine Na^+

Cl^-

OH^-

Sodium hydroxide solution

The membrane process is being used more and more. Although it is expensive to set up and run, its products are very pure, and it does not involve the poisonous metal mercury.

In the diaphragm process, a barrier (the diaphragm) is used to keep the products from mixing and reacting with each other. This method produces rather impure products that are suitable only for some uses. The mercury process has one electrode made from liquid mercury, and it produces pure products. The membrane process is similar to the diaphragm process but uses purer ingredients and a more complex material for the barrier.

Special characteristics

DANGER FACTS

Chlorine gas is a very dangerous substance. These are the lowest concentrations of chlorine that will cause the effect listed:

- detectable by smell: three parts in a million

- causes coughing: three parts in 100,000

- causes poisoning after several hours: one part in a million

- causes death after a few deep breaths: one part in 1,000.

Chlorine is easily recognizable in its natural state. It is a yellow-green gas, more than twice as heavy as air, with an intense, pungent smell. The fact that it is so abundant on Earth and so reactive makes chlorine extremely useful.

Although it is one of the few elements that is a gas at room temperature, chlorine is usually supplied to laboratories and industrial plants as a liquid. This is because chlorine needs to be cooled to only −29°F (−34°C) to liquefy—compared with −297°F (−183°C) for oxygen and −423°F (−253°C) for hydrogen. Chlorine is not as colorful when it is a liquid as when it is a gas, and it is even paler as a solid (it freezes at −150°F, or −101°C).

Careful handling

You are unlikely to have encountered chlorine as an element—which is a good thing, because chlorine gas is very poisonous and can kill at very low concentrations. This, together with its extremely reactive nature, means that chlorine must be handled very carefully. People dealing with chlorine must always have a gas mask ready, in case any gas escapes accidentally. Chlorine must also be kept dry when being transported or stored, because it forms a concentrated acid when it mixes with water, and this would attack the metal container.

Chlorine is seen in its pure gaseous state only in laboratories and industrial plants, after it has been isolated from its natural compounds. This highly toxic gas irritates the eyes, nose, throat, and lungs.

How chlorine reacts

When the metal, sodium, is placed in a flask of chlorine gas, the resulting reaction is highly vigorous. The product of this reaction is the stable compound, sodium chloride, or common salt.

Chlorine forms an amazing array of compounds and reacts with almost every other element in the periodic table. It can even form compounds with gold. This feature of chlorine is very unusual, because gold is famous for being the least reactive of all the metals.

Like other halogens, chlorine reacts with elements to fill the space in its outer electron shell. There are two ways a chlorine atom can do this. It can snatch an electron away from another atom, forming separate charged atoms called ions, or it can "share" an electron with another atom, creating a covalent bond between them and forming a chlorine compound.

Chlorine compounds

Some chlorine compounds are very safe and stable indeed. Atoms with just one electron in their outer shell are natural partners for chlorine atoms. For example, the highly reactive metal, sodium, happily gives away its electron to chlorine, and together they form the sodium and chloride ions in the compound, sodium chloride. Also known as table salt, this compound is safe enough to eat.

Chlorine can also form compounds that are very reactive, however. For example, when chlorine combines with hydrogen,

the hydrogen shares its one electron with chlorine—this sharing of electrons is known as a covalent bond—completing the shell and forming hydrogen chloride. When hydrogen chloride gas dissolves in water, it becomes hydrochloric acid. The hydrogen atom then leaves its electron with the chlorine, and hydrogen and chloride ions are created. These ions wander off among the water molecules, free to react with other substances they meet. Among other things, hydrochloric acid will react with most metals—even those we think of as resistant to corrosion.

When the chloride ions in hydrochloric acid attack metals, a compound is formed between the chlorine and the metal. For example, aluminum and silver react with chlorine to form aluminum chloride and silver chloride, respectively. These compounds are ionic, because—like sodium—metals are inclined to give up

Concentrated hydrochloric acid can even attack stainless steel, so it is normally stored in glass or plastic containers. The acid in this glass beaker is attacking chips of limestone, a rock made up of calcium carbonate (CaCO$_3$).

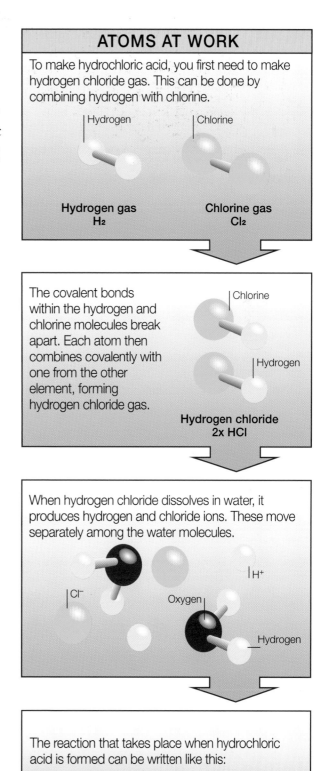

ATOMS AT WORK

To make hydrochloric acid, you first need to make hydrogen chloride gas. This can be done by combining hydrogen with chlorine.

Hydrogen

Chlorine

Hydrogen gas
H$_2$

Chlorine gas
Cl$_2$

The covalent bonds within the hydrogen and chlorine molecules break apart. Each atom then combines covalently with one from the other element, forming hydrogen chloride gas.

Chlorine

Hydrogen

Hydrogen chloride
2x HCl

When hydrogen chloride dissolves in water, it produces hydrogen and chloride ions. These move separately among the water molecules.

H$^+$

Cl$^-$

Oxygen

Hydrogen

The reaction that takes place when hydrochloric acid is formed can be written like this:

H$_2$ + Cl$_2$ → 2HCl → 2H$^+$ + 2Cl$^-$

Some perchlorate compounds are used as propellants in rocket engines. When these are lit, an explosive chemical reaction thrusts the rocket into space.

their distant outer electrons rather than insist on sharing an electron with chlorine.

As you might expect, chlorine also reacts with nonmetallic elements such as sulfur, phosphorus, and oxygen. Unlike the metals, most of these react with chlorine to form covalent compounds. With oxygen, chlorine forms a variety of covalent compounds. For example, Cl_2O, Cl_2O_3,

DID YOU KNOW?

EXPLODING COMPOUNDS

Some oxygen-rich chlorine compounds are highly explosive. Perchloric acid, which contains four atoms of oxygen for every atom of chlorine, explodes in contact with organic materials, and sometimes on its own. A related chlorine compound, potassium perchlorate, is used in matches. Another perchlorate compound— sodium perchlorate—is combined with powdered aluminum to provide the explosions that propel the space shuttle forward at immense speed in an oxygen-free atmosphere.

Chlorine trifluoride can set fire to substances such as wood. It was used in German incendiary bombs in World War II, causing scenes of devastation.

ClO_2, Cl_2O_6, Cl_2O_4, and Cl_2O_7 are all known, although these compounds are generally very unstable. Similarly, chlorine combines with its halogen neighbor, fluorine, to form chlorine trifluoride (ClF_3), which is one of the most reactive substances known to exist.

Organic compounds

Chlorine also forms a huge range of different organic compounds—compounds that contain several carbon atoms linked together. Plastics, oils, and the substances in your body are all organic compounds. When chlorine joins an organic compound, it often replaces a hydrogen atom in a substitution reaction. Several hydrogen atoms can be replaced in the same molecule, one after the other, adding to the variety of compounds produced.

ATOMS AT WORK

Hydrochloric acid reacts with metals to form metal chlorides. If aluminum metal is mixed with hydrochloric acid, it is attacked by the hydrogen and chloride ions in the acid.

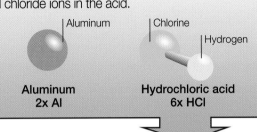

Aluminum
2x Al

Hydrochloric acid
6x HCl

Each aluminum atom in the metal gives three electrons to the hydrogen ions in the acid. This produces hydrogen atoms, which form hydrogen gas. The gas then bubbles out of the solution, leaving the aluminum atoms as positively charged ions.

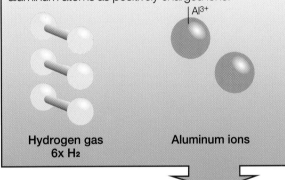

Hydrogen gas
6x H_2

Aluminum ions

The aluminum ions move into the solution with the chloride ions to form aluminum chloride. Because they are in solution, the ions are separate.

Aluminum chloride
2x $AlCl_3$

The reaction that takes place when hydrochloric acid reacts with aluminum can be written like this:

$$2Al + 6HCl \rightarrow 2AlCl_3 + 3H_2$$

Chlorine in nature

Chlorine is abundant in nature: in living things as well as in the earth and in the oceans. Many plants and animals make chlorine compounds and use them in their body processes. In fact, over 1,000 different compounds containing chlorine are known to occur in nature.

Chlorine in marine organisms

The ocean is where most of the world's chlorine is found, and sea plants and animals are especially good at producing chlorine compounds. For example, chloroform and tetrachloromethane are both produced by seaweeds. On land, wood-rotting fungi produce large amounts of chloromethane, while some plants and animals use chlorine compounds to protect themselves from predators. For example, a species of tree frog from Ecuador produces a complex chlorine compound that seems to dissuade birds and other predators from attacking it. This compound appears to have painkilling properties in humans, and it is said to be

The chlorine compounds that occur naturally in seaweeds are rather smelly and are probably responsible for the seaside's characteristic smell.

several times more powerful than morphine (a very strong painkiller). Other animals use chlorine compounds to communicate—the lone star tick excretes the chlorine compound, dichlorophenol, as an odor, or pheromone, to attract a mate.

Chlorine in the body

In humans and other mammals, chlorine is a truly vital element. We take in more chlorine than any other mineral, mostly in the form of salt. For adults, this is ⁹⁄₁₀ of an ounce (6 grams) per day—about six times the amount of calcium and 400 times the amount of iron we take in. Much of this chlorine passes out of the body again. If this happens too quickly, however—for example, if someone is vomiting, has diarrhea, or is sweating a lot in hot weather—chlorine levels can become dangerously low.

Chloride ions circulate in our blood, and they are regulated to help keep the concentration of fluids in our body at the right level. Two thirds of the negative ions in the blood are chloride ions, although the concentration inside cells is much lower. Our stomachs use hydrochloric acid to help break down proteins in our food into smaller molecules, and also to kill off harmful bacteria before they are able to travel farther into the body.

This magnified picture shows a white blood cell (brown) destroying harmful bacteria (yellow). These cells, called macrophages, use chlorine in their attack.

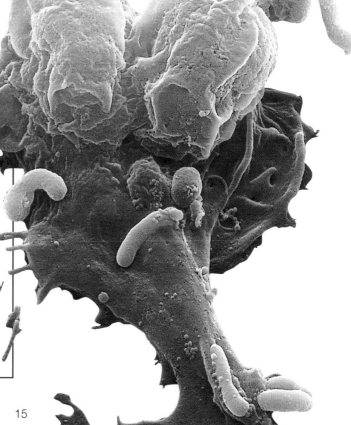

DID YOU KNOW?

ATTACKING INVADERS

According to recent research, the cells in the immune system use chlorine to attack harmful invaders, such as bacteria. Enzymes in white blood cells are able to convert some of the chloride ions circulating in the blood to hypochlorite ions. These attack the germs directly or help to activate other substances to make an attack. Chlorine thus helps white blood cells to kill off germs and keep us healthy.

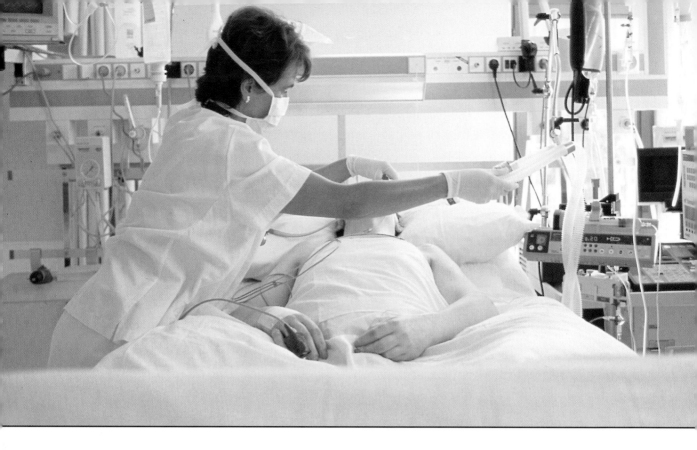

Disinfectants and bleach

Cleanliness is particularly important in the intensive care unit of a hospital. Many of the disinfectants used to keep this area germ-free contain chlorine.

Chlorine has an important role in keeping us healthy and our surroundings clean. Its compounds have long been used as disinfectants—in the home, in hospitals, and anywhere else where hygiene is important.

Killing germs

Chlorine kills germs, such as bacteria and viruses, by interfering with the chemistry of the molecules inside them. Chlorine gas is not normally used for this purpose. Instead, some disinfectants contain chlorine compounds that are capable of transferring their chlorine atoms to other compounds, such as the enzymes within bacteria and other cells.

Enzymes are protein molecules that carry out all the essential tasks within a cell, and each one needs to be a particular shape to do its job. When an enzyme is attacked by chlorine, one or more hydrogen atoms in the protein molecule is replaced by chlorine—and this can be enough to make the whole molecule change its shape or break apart completely. Without properly functioning enzymes, the cell or bacterium dies.

Chlorine as bleach

Hypochlorous acid, a solution that works as bleach, can be made simply by bubbling chlorine through water. This acid is called an active chlorine compound, because it contains a chlorine atom that easily reacts with organic molecules and so acts as a disinfectant. The solution is chemically unstable, however, and chlorine gas

Many disinfectants in use today are based on chlorine compounds. For example, TCP is a strong-smelling disinfectant used in some countries for treating cuts and scrapes in the home. It gets its name because it contains the chlorine compound trichlorophenol. An even more complex-sounding chlorine compound, sodium dichloroisocyanurate, is sold under different brand names and is used for sterilizing babies' bottles and for keeping swimming pools clean. As an all-purpose disinfectant for surfaces and surroundings, however, the most important substance is bleach.

Today, drug cabinets in many homes contain chlorine-based disinfectants. These sting when they are applied to cut skin, but they help to clean the wound and prevent it from becoming infected.

gradually escapes from it. This makes it dangerous to use—and it also makes it ineffective after a while, because it is constantly losing its chlorine to the air.

The problem of chemical instability has been solved in the bleach you can buy in bottles. This bleach is made by bubbling chlorine gas through an alkali solution, usually sodium hydroxide. Chlorine reacts with the sodium hydroxide to form sodium hypochlorite, which is just as effective a disinfectant as hypochlorous acid but is chemically stable.

French chemist Claude-Louis Berthollet, the inventor of bleach, was made a senator and a count by Napoleon Bonaparte.

Household bleach is safe to use, therefore, but only if you do not mix it with anything acidic. If this happens, the hypochlorite will become unstable again and poisonous chlorine gas may be produced. For this reason, bleach bottles always carry warnings not to use bleach with any other cleaning product, as many of these products are acids.

DID YOU KNOW?

BERTHOLLET'S BLEACH

Bleach is not a new idea. It was first marketed as "Eau de Javelle" in 1789, just 15 years after chlorine was discovered. The inventor of this product was French chemist Claude-Louis Berthollet (1748–1822), who realized in 1785 that a solution of chlorine in the alkali potassium hydroxide could be used to lighten the color of fabrics. Berthollet's invention was a great success with the clothmaking industry, because, before that, the only way of bleaching fabrics was to lay them out in the Sun for several weeks.

Chlorine in drinking water

The germ-killing effect of chlorine has also saved huge numbers of lives by keeping water supplies clean. Each year, around three million people worldwide still die as a result of waterborne diseases, including typhoid, cholera, and dysentery. All these diseases can be prevented by ensuring clean water supplies.

Chlorine was first used to sterilize water supplies in London, Britain, in 1850 after an outbreak of cholera there. Regularly adding chlorine to water supplies began in the rest of the country shortly after

1900, and this greatly reduced deaths from typhoid. Towns and cities in the United States soon began to do the same. This was so effective that the number of cases of cholera in the U.S. fell from 25,000 in 1900 to just 20 in 1960. Over the same period, the average life span in the whole of the developed world has increased by around 50 percent—from about 50 years to 75 years. Adding chlorine to water has played a major part in achieving this.

Today, other ways of producing clean drinking water are being developed, but chlorine is still used in most water treatment systems. While some people are concerned about possible side effects from

This cartoon, published in 1849, was in protest to the filthy water supply in London, which caused an outbreak of the waterborne disease cholera.

drinking water containing chlorine, most scientists believe it is quite safe. Many people do not like the taste of chlorine, but it can be easily removed from your faucet water by a water filter.

Chlorine gas is added directly to water before it arrives at your home. When the water comes out of your faucets, it probably has a chlorine concentration of less than one part in a million (0.0001 percent).

Chlorine in medicine

When chlorine was discovered, it was still not widely known that germs such as bacteria and viruses cause and spread disease. By the early 19th century, however, some doctors began to realize that many diseases are contagious, and their spread might be limited by better cleanliness in hospitals.

Soon, doctors started to experiment with chlorine-based chemicals as disinfectants. For example, in 1835, U.S. doctor and writer Oliver Wendell Holmes (1809–1894) began insisting that midwives washed their hands in chloride of lime (calcium hypochlorite) solution before delivering babies. While this dramatically reduced the number of women dying from infection after giving birth, the medical profession at the time opposed this practice, saying Holmes's ideas were an insulting criticism of the way they worked.

The chlorine-containing antibiotic vanomycin is used to treat infections such as pneumonia caused by Staphylococcus *bacteria, shown below.*

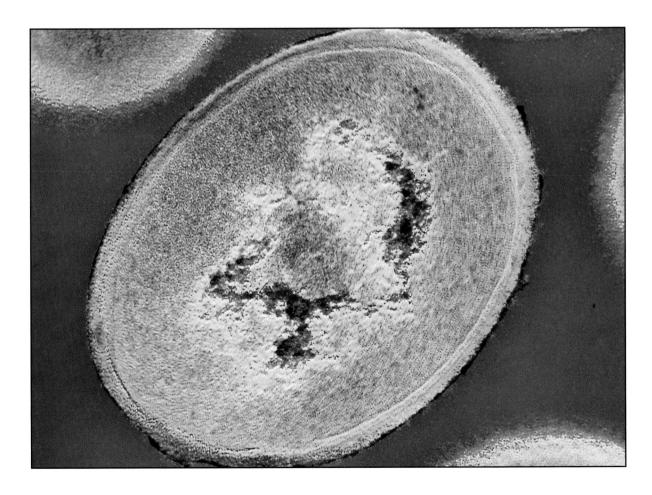

It was only later in the century, after French scientist Louis Pasteur (1822–1895) had demonstrated that disease was spread by tiny organisms, that the use of disinfectants such as chlorine bleach became routine.

Chlorine in drugs

Disinfectants are still extremely important in medical care today, but much more complex chemicals are also used to kill germs and cure diseases within the body. Around 30 percent of modern medicines contain chlorine compounds of some sort. In particular, the powerful antibiotic, vanomycin, which kills a type of bacterium, called *Staphylococcus*, is a chlorine compound. The drug, chloroquine, which is used to treat malaria, is also a chlorine compound. Some herbal drugs contain chlorine—for example, one plant used in traditional Chinese medicine contains five different chlorine compounds.

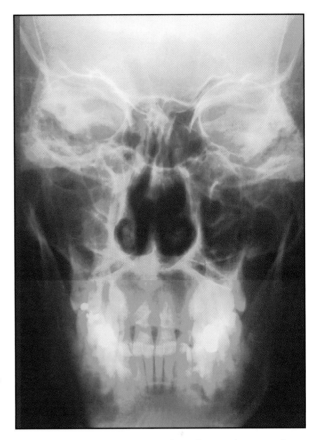

Silver chloride has been used in the production of X-ray plates ever since the discovery of X rays by German physicist Wilhelm Röntgen in 1895.

Chlorine also plays a part in medical technology. While most photographic film is made using silver bromide or silver iodide, the "films" used in X-ray images use silver chloride. When X rays strike the silver chloride molecules, they cause a reaction in which silver metal is formed. When the X-ray plate is developed, the places where the silver metal is present show up as dark areas, producing the characteristic negative image that you see in X-ray photographs.

DID YOU KNOW?

SENDING PEOPLE TO SLEEP

The first anesthetic to be used in surgery was a chlorine compound. First used in 1847, chloroform ($CHCl_3$) is very powerful, and it remained in use for nearly 100 years. Chloroform is not used as an anesthetic today, however, because we now know that it can cause damage to body tissues, especially the liver.

Chlorine in industry

The first suggested use for the plastic PVC was in making shower curtains. Now it is made into many products, including tough waterproof clothing.

Chlorine is one of the basic materials of the chemical industry, and the range of products that is manufactured using this element is truly enormous.

Chlorine's uses

About 65 percent of the chlorine used in industry goes to produce organic chemicals, including plastics, while 20 percent is used to make bleach and disinfectants. The rest is used to make inorganic (non-carbon containing) compounds between chlorine and other elements, such as iron, zinc, and titanium.

The 10,000 or so different chlorine compounds used in manufacturing are produced from a smaller number of chlorine-containing chemicals, many of which are organic compounds. For example, dichloroethene (CH_2Cl_2) is used to make more complex compounds such as propylene oxide and polypropylene glycol. These are used to make the substances that actually go into products. Propylene oxide is turned into the plastic polyurethane, which is used in paints, foam cushions, and car bumpers, while polypropylene glycol is used to make food additives, pesticides, and antifreeze solutions (solutions that lower the freezing point of water).

One of the most useful chlorine products is PVC. This plastic is made into pipes and tubing, cables and wire insulation, floorings and windows, bottles, adhesives, and waterproof clothing—to name just a few of its many uses. Like all plastics, PVC is made by chemically linking together many smaller molecules, or monomers, to make a much larger molecule, called a polymer.

Bleach in industry

As well as in the home and in hospitals, chlorine bleaches are used in the paper industry. However, this industry is now trying to find ways to reduce the amount of waste chlorine flowing out into rivers and the environment, because it can

In the paper industry, chlorine bleaches are used in huge quantities to help turn wood into white paper.

combine with naturally occurring compounds to turn them into poisons. Other industries are also under pressure to use less chlorine, so while chlorine is a vitally important industrial chemical now, it may be less used in years to come.

Chlorine in war

S adly, chlorine gas is a poisonous substance that has been used to kill and injure people in war. It was first used on April 22, 1915, by German troops during World War I (1914–1918).

Blanket of gas

Chlorine is so much heavier than air that it forms a blanket of gas close to the ground. When it was used during World War I, it flowed down into the trenches where the soldiers were hiding. This gas attacks the lungs, making them fill with fluid so that the person breathing it "drowns" from the inside. It also attacks the nose, throat, and eyes. Many thousands of tons of chlorine gas were used in World War I, with devastating effects.

This newspaper reports the use of chlorine gas during World War I. The deadly gas was said to smell like pepper and pineapple mixed together.

Phosgene and mustard gas

Soon after the first chlorine attack in 1915, both sides in World War I began developing compounds that were even more lethal than chlorine gas—and many of these also contained chlorine.

One example is phosgene (Cl_2O), which also causes severe injury to the lungs.

Mustard gas, which was used in large quantities by both sides, caused more injuries than any other chemical in World War I. This substance burns and produces

World War I was the first war in which chemical warfare had been used. Gas masks were introduced and were issued to the horses as well as the soldiers.

blisters on the skin and in the lungs, which can become infected. Lewisite, which contains arsenic as well as chlorine, is particularly lethal and can cause damage to body tissues in concentrations of less than one part in 1,000 million. Although lewisite was developed in the U.S. for use during World War I, it was

never actually used—the first lewisite weapons were still being made when the war ended and peace was declared.

Chemical weapons ban

Today, the use of chemical weapons is forbidden. However, some countries may be producing and storing them in secret, and the United Nations (the international organization that tries to keep peace between countries) has been only partly successful in controlling this activity.

DID YOU KNOW?

WAR POETRY

British poet Wilfred Owen (1893–1918) fought and was killed in World War I. In the second verse of a poem called *"Dulce et Decorum est"* (Latin for "It is Sweet"), he wrote about his experience of a chlorine attack and its effects:

> Gas! GAS! Quick, boys!—An ecstasy of fumbling,
> Fitting the clumsy helmets just in time;
> But someone still was yelling out and stumbling
> And flound'ring like a man in fire or lime. —
> Dim, through the misty panes and thick green light,
> As under a green sea, I saw him drowning.
>
> In all my dreams, before my helpless sight,
> He plunges at me, guttering, choking, drowning.

The title is taken from the last two lines of this poem, where Owen says that it is an "old lie" to say that "It is sweet to die for your country" ("Dulce et decorum est Pro patria mori"). By saying this, Owen shows his contempt for war.

Chlorine in the environment

Chlorine compounds are used so much in industry because they are extremely stable and do not break down very easily. Unfortunately, if they are released into the environment the long-lasting character of chlorine compounds can cause a real problem for wildlife. This is because they get taken into the bodies of animals, which are then eaten by other animals until the

The bald eagle, which is the national symbol of the United States, has been removed from the endangered list since the banning of DDT in 1972.

compounds become highly concentrated in predators at the top of the food chain. For example, this is currently occurring with poisonous chlorine substances, called dioxins, which are produced by many industrial processes. It is thought that dioxins can cause cancer and babies to be born deformed, even when the concentration of dioxins in the environment is low.

Ozone layer under attack

Chlorine is also very reactive, so any chlorine that gets into the environment is likely to form all kinds of compounds whose effects cannot be predicted. In the upper atmosphere, chlorine compounds called CFCs have damaged the ozone layer, which protects us from harmful ultraviolet (UV) rays in sunlight. Because

of international cooperation to stop this damage, CFCs—which were used in refrigerators and aerosol sprays—are no longer produced in the U.S. and Europe.

Some people are now calling for a complete ban on the use of chlorine and volatile (easily turned to gas) chlorine compounds. This is unlikely to happen in the near future, but governments are trying more closely to control the amount of chlorine entering the environment. Work is being put into finding less polluting ways of using chlorine and its compounds—or finding chlorine-free alternatives.

Because of concerns about the impact of chlorine and chlorine compounds on the environment, the Olympics 2000 stadium in Sydney (shown below) was built without using any PVC.

Periodic table

Everything in the Universe is made from combinations of substances called elements. Elements are the building blocks of matter. They are made of tiny atoms, which are much too small to see.

The character of an atom depends on how many even tinier particles called protons there are in its center, or nucleus. An element's atomic number is the same as the number of protons.

Scientists have found more than 110 different elements. About 90 elements occur naturally on Earth. The rest have been made in experiments.

All these elements are set out on a chart called the periodic table. This lists all the elements in order according to their atomic number.

The elements at the left of the table are metals. Those at the right are nonmetals. Between the metals and the nonmetals are the metalloids, which sometimes act like metals and sometimes like nonmetals.

○ On the left of the table are the alkali metals. These elements have just one electron in their outer shells.

○ On the right of the periodic table are the noble gases. These elements have full outer shells.

○ Elements in the same group have the same number of electrons in their outer shells.

○ Elements get more reactive as you go down a group.

○ The number of electrons orbiting the nucleus increases down each group.

○ The transition metals are in the middle of the table, between Groups II and III.

Transition metals

Group I	Group II							
1 **H** Hydrogen 1								
3 **Li** Lithium 7	4 **Be** Beryllium 9							
11 **Na** Sodium 23	12 **Mg** Magnesium 24							
19 **K** Potassium 39	20 **Ca** Calcium 40	21 **Sc** Scandium 45	22 **Ti** Titanium 48	23 **V** Vanadium 51	24 **Cr** Chromium 52	25 **Mn** Manganese 55	26 **Fe** Iron 56	27 **Co** Cobalt 59
37 **Rb** Rubidium 85	38 **Sr** Strontium 88	39 **Y** Yttrium 89	40 **Zr** Zirconium 91	41 **Nb** Niobium 93	42 **Mo** Molybdenum 96	43 **Tc** Technetium (98)	44 **Ru** Ruthenium 101	45 **Rh** Rhodium 103
55 **Cs** Cesium 133	56 **Ba** Barium 137	71 **Lu** Lutetium 175	72 **Hf** Hafnium 179	73 **Ta** Tantalum 181	74 **W** Tungsten 184	75 **Re** Rhenium 186	76 **Os** Osmium 190	77 **Ir** Iridium 192
87 **Fr** Francium 223	88 **Ra** Radium 226	103 **Lr** Lawrencium (260)	104 **Unq** Unnilquadium (261)	105 **Unp** Unnilpentium (262)	106 **Unh** Unnilhexium (263)	107 **Uns** Unnilseptium (?)	108 **Uno** Unniloctium (?)	109 **Une** Unnilennium (?)

Lanthanide elements

Actinide elements

57 **La** Lanthanum 39	58 **Ce** Cerium 140	59 **Pr** Praseodymium 141	60 **Nd** Neodymium 144	61 **Pm** Promethium (145)
89 **Ac** Actinium 227	90 **Th** Thorium 232	91 **Pa** Protactinium 231	92 **U** Uranium 238	93 **Np** Neptunium (237)

The horizontal rows are called periods. As you go across a period, the atomic number increases by one from each element to the next. The vertical columns are called groups. Elements get heavier as you go down a group. All the elements in a group have the same number of electrons in their outer shells. This means they react in similar ways.

The transition metals fall between Groups II and III. Their electron shells fill up in an unusual way. The lanthanide elements and the actinide elements are set apart from the main table to make it easier to read. All the lanthanide elements and the actinide elements are quite rare.

Chlorine in the table

Chlorine has 17 protons in its nucleus, so it has atomic number 17. This element is one of the halogens in Group VII of the periodic table. Like all the other halogens, such as fluorine, it has seven electrons in its outer shell. It readily undergoes chemical reactions with other elements to form a large number of compounds.

Metals
Metalloids (semimetals)
Nonmetals

17
Cl
Chlorine
35

Atomic (proton) number
Symbol
Name
Atomic mass

Group VIII

					Group III	Group IV	Group V	Group VI	Group VII	2 He Helium 4
					5 B Boron 11	6 C Carbon 12	7 N Nitrogen 14	8 O Oxygen 16	9 F Fluorine 19	10 Ne Neon 20
					13 Al Aluminum 27	14 Si Silicon 28	15 P Phosphorus 31	16 S Sulfur 32	17 Cl Chlorine 35	18 Ar Argon 40
28 Ni Nickel 59	29 Cu Copper 64	30 Zn Zinc 65			31 Ga Gallium 70	32 Ge Germanium 73	33 As Arsenic 75	34 Se Selenium 79	35 Br Bromine 80	36 Kr Krypton 84
46 Pd Palladium 106	47 Ag Silver 108	48 Cd Cadmium 112			49 In Indium 115	50 Sn Tin 119	51 Sb Antimony 122	52 Te Tellurium 128	53 I Iodine 127	54 Xe Xenon 131
78 Pt Platinum 195	79 Au Gold 197	80 Hg Mercury 201			81 Tl Thallium 204	82 Pb Lead 207	83 Bi Bismuth 209	84 Po Polonium (209)	85 At Astatine (210)	86 Rn Radon (222)

62 Sm Samarium 150	63 Eu Europium 152	64 Gd Gadolinium 157	65 Tb Terbium 159	66 Dy Dysprosium 163	67 Ho Holmium 165	68 Er Erbium 167	69 Tm Thulium 169	70 Yb Ytterbium 173
94 Pu Plutonium (244)	95 Am Americium (243)	96 Cm Curium (247)	97 Bk Berkelium (247)	98 Cf Californium (251)	99 Es Einsteinium (252)	100 Fm Fermium (257)	101 Md Mendelevium (258)	102 No Nobelium (259)

Chemical reactions

Chemical reactions are going on all the time—candles burn, nails rust, food is digested. Some reactions involve just two substances; others many more. But whenever a reaction takes place, at least one substance is changed.

In a chemical reaction, the atoms stay the same. But they join up in different combinations to form new molecules.

Writing an equation

Chemical reactions can be described by writing down the atoms and molecules before and the atoms and molecules after. Since the atoms stay the same, the number of atoms before will be the same

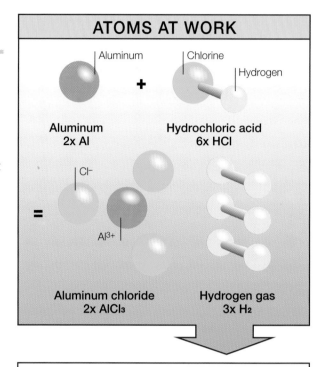

ATOMS AT WORK

Aluminum

Chlorine

Hydrogen

+

Aluminum
2x Al

Hydrochloric acid
6x HCl

Cl⁻

=

Al³⁺

Aluminum chloride
2x AlCl₃

Hydrogen gas
3x H₂

The reaction that takes place when hydrochloric acid reacts with aluminum can be written like this:

$$2Al + 6HCl \rightarrow 2AlCl_3 + 3H_2$$

as the number of atoms after. Chemists write the reaction as an equation. The equation shows what happens in the chemical reaction.

When the numbers of each atom on both sides of the equation are equal, the equation is balanced. If the numbers are not equal, something is wrong. The chemist adjusts the number of atoms involved until the equation does balance.

Here scientists are developing a process for creating aluminum chloride by dissolving an aluminum-containing clay called kaolin in hydrochloric acid.

Glossary

acid: A substance that can provide hydrogen atoms for chemical reactions.

anesthetic: A substance that is used to put a person to sleep or to numb an area of the body before surgery.

antibiotic: A drug that kills or prevents the growth of bacteria.

atom: The smallest part of an element that has all the properties of that element.

atomic mass: The number of protons and neutrons in an atom.

bond: The attraction between two atoms that holds them together.

compound: A substance made of atoms of more than one element. The atoms are held together by chemical bonds.

corrosion: The eating away of a material by reaction with other chemicals, often oxygen and moisture in the air.

electrode: A material through which an electrical current flows into, or out of, a liquid called an electrolyte.

electrolysis: The use of electricity to change a substance chemically.

electron: A tiny particle with a negative charge. Electrons are found inside atoms, where they move around the nucleus in layers called electron shells.

element: A substance that is made from only one type of atom. Chlorine is one of the elements called the halogens.

enzyme: A protein molecule that has a specific function within a living cell.

ion: A particle that is similar to an atom but that carries a negative or positive electrical charge.

isotopes: Atoms of an element with the same number of protons and electrons but different numbers of neutrons.

molecule: A particle that contains atoms held together by chemical bonds.

neutron: A tiny particle with no electrical charge. It is found in the nucleus of almost every atom.

nonmetal: An element on the right-hand side of the periodic table.

nucleus: The center of an atom. It contains protons and neutrons.

organic compound: A compound that contains a chain of carbon atoms.

ozone layer: A layer of the atmosphere that protects Earth from ultraviolet rays.

periodic table: A chart of all the chemical elements laid out in order of their atomic number.

proton: A tiny particle with a positive charge. Protons are found inside the nucleus of an atom.

solvent: A liquid that can dissolve one or more other substances.

substitution reaction: A reaction in which one atom or part of a molecule is replaced by another.

ultraviolet: A form of radiation similar to light but invisible to the naked eye.

Index